I0453104

Copyright © 2023

All rights reserved. No part of this book may be reproduced or used in any manner without the prior written permission of the copyright owner, except for the use of brief quotations in a book review.

Tech Baby LLC
https://thetechbaby.com

This book is dedicated to Verona, Angelo & Arianna.

Volume I

Tech Baby LLC
https://thetechbaby.com

Illustrations by z3n

ABC's

for

tech babies

By Meg Gonzalez

A is for

Anti-Virus

A program that scans for
harmful programs.

B is for

Bit
**The tiniest piece of
information in a computer.**

C is for

Cryptocurrency
A type of money that is only electronic.

D is for

Data
Information a computer
can save, send, or use.

E is for

Encryption
Protection of data with a
special code.

F
is for

Firewall

A device or program that protects your computer on the Internet.

G is for

Gigabyte
The data equal to 1,024 megabytes.

H is for

Hacker

A person that can steal
information from an
electronic device.

I is for

Input
To bring data into your computer.

J is for

Java
A type of language that computers understand.

K is for

Kilobyte
The data equal to 1,024 bytes.

L is for

Linux

Free software that can run
your computer.

M is for

Malware
A bad program made to
harm your computer.

N is for

Network
Connected computers that can share data.

O is for

Output
To print or display information from a computer.

P is for

Password
A secret code that allows you to use a program or computer.

 is for

Query

To ask a computer for information.

R is for

Re-image
To erase all programs and start over.

S is for

Server
A powerful computer that
provides information to
other computers.

T is for

Terabyte
The data equal to 1,024 gigabytes.

U
is for

Upload
Transfer of data from one
computer to another.

VPN

A safe way to connect to a
network.

W
is for

Wi-Fi

A connection to the
Internet without wires.

 is for

XML

**A language used to share
information on the Internet.**

Y

is for

Yottabyte
The data equal to a trillion
terabytes (yes, a lot!).

Z is for

Zero-day

A big problem in a
computer that is new and
hasn't been fixed yet.

www.ingramcontent.com/pod-product-compliance
Lightning Source LLC
Chambersburg PA
CBHW041528120626
46551CB00018B/2614